Collins

Counting bumper book

Ages 3–5

1 5 4 6 2 7 3

7

Carol Medcalf

How to use this book

- Find a quiet, comfortable place to work, away from distractions.

- This book has been written in a logical order, so start at the first page and work your way through.

- Try to use the following language as you work through the book together: add, plus, take away, minus, more than, less than, equals, all together, left over, set(s), half.

- Help with reading the instructions where necessary and ensure that your child understands what to do.

- When questions have two parts, it is often best to gain the first answer and record it before moving on to the next part of the question.

- All children learn and develop at a different rate. If an activity is too difficult for your child, then do more of our suggested practical activities (see Parent's tips) and return to the page when you know that they are likely to achieve it.

- Some children find it easier if all the other activities on the page are covered with a blank piece of paper, so only the activity they are working on is visible.

- Always end each activity before your child gets tired so that they will be eager to return next time.

- Help and encourage your child to check their own answers as they complete each activity.

- Let your child return to their favourite pages once they have been completed. Talk about the activities they enjoyed and what they have learned.

Special features of this book:

- **Parent's tip:** situated on every left-hand page, this suggests further activities and encourages discussion about what your child has learned.

- **Progress panel:** situated at the bottom of every right-hand page, the number of animals and stars shows your child how far they have progressed through the book. Once they have completed each double page, ask them to colour in the blank star.

- **Certificate:** the certificate on the last page should be used to reward your child for their effort and achievement. Remember to give your child plenty of praise and encouragement, regardless of how they do.

Published by Collins
An imprint of HarperCollins*Publishers* Ltd
The News Building
1 London Bridge Street
London
SE1 9GF

Browse the complete Collins catalogue at www.collins.co.uk

© HarperCollins*Publishers* Ltd 2018

10 9 8 7 6 5 4 3

ISBN 9780008275457

The author asserts the moral right to be identified as the author of this work.

British Library Cataloguing in Publication Data

A Catalogue record for this publication is available from the British Library

All images and illustrations are © Shutterstock.com and © HarperCollins*Publishers*

Author: Carol Medcalf
Commissioning Editor: Michelle I'Anson
Project Manager: Rebecca Skinner
Cover Design: Sarah Duxbury
Text Design and Layout: QBS Learning
Illustration: Jenny Tulip
Production: Natalia Rebow
Printed by Martins the Printers

Contents

Writing 0–10

● Start at the red dot. Join the dots to write the number. Count the green counters.

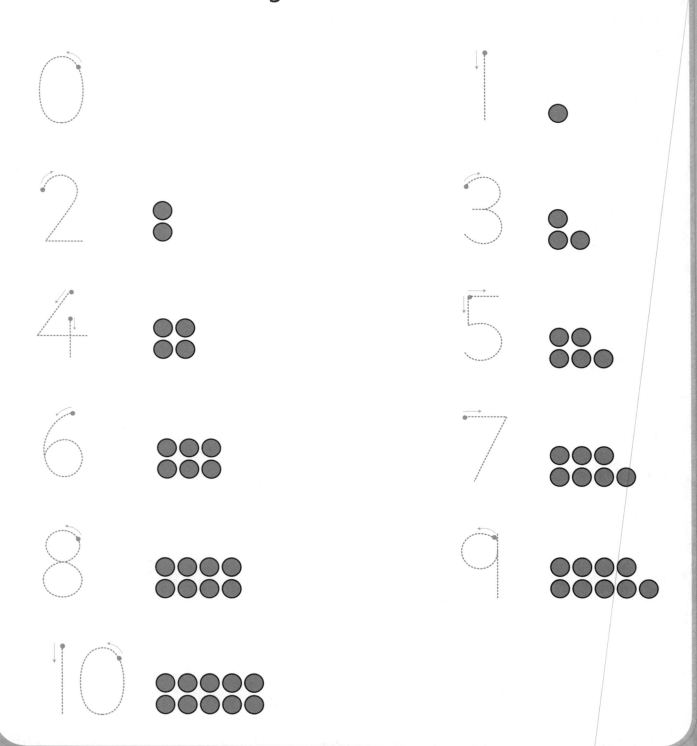

As there will be lots of activities in this book that require your child to write a number, this page is ideal for practising the correct way. For further practice, you can write numbers in pencil or a light-coloured pen for your child to trace.

Writing 11–20

Start at the red dot. Join the dots to write the number. Count the green counters.

Counting 0–5

- Count how many animals are in each field.
  Draw a line from the picture to the correct number

0   1   2   3   4   5

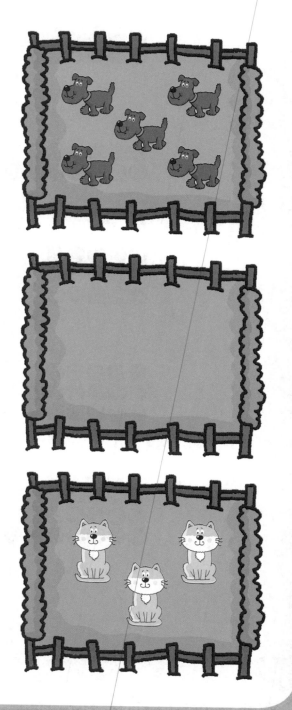

More counting 0–5

- Draw a line to match each mole to a molehill.
  Did every molehill have a mole?

- The rabbits want some carrots.
  Draw the correct number of carrots  in the
  box below each rabbit.

| 0 | 1 | 2 | 3 | 4 | 5 |

Counting 6

- Draw **6** wheels on the train.

- Count the stars on the hot air balloon. Fill each one in a different colour.

Counting 7

- Draw a circle around the group that has **7** cars.

- Draw **7** people on the bus.

Well done!
Now colour
the star.

Counting 8

- Count and colour **8** cacti.

- Count the flowers.
Draw a line from each bunch of **8** flowers to the big number **8**.

Make a daisy chain with your child. Count how many flowers you pick. Then count how many you need to make the flowers into a chain for a bracelet or necklace.

Counting 9

● Count the flowers.
Tick (✓) the garden that has **9** flowers growing.

● How many daisies are in the daisy chain?

Well done!
Now colour
the star.

Counting 10

● Count how many boats there are.
Write the number **10** under each boat.

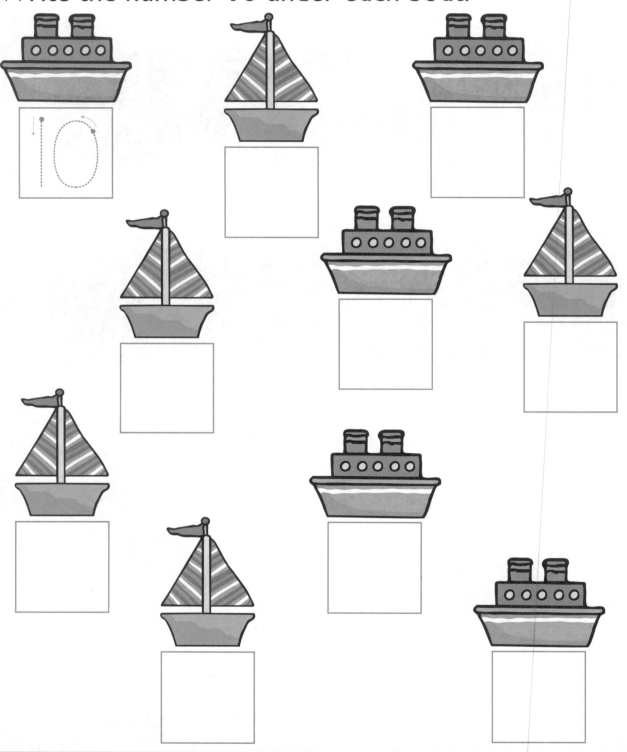

Counting puzzles

Count the number of dots on each puzzle piece. Draw a line to match it to the piece with the correct number.

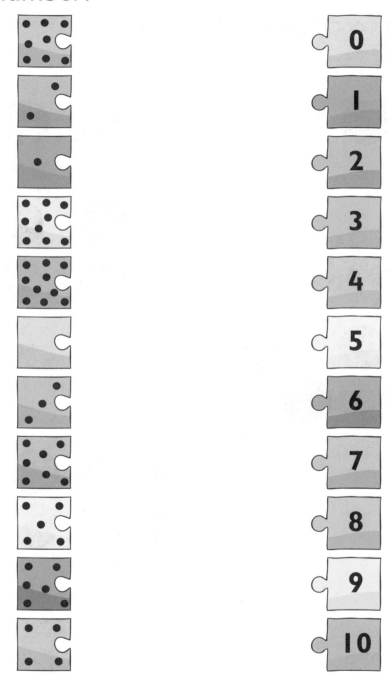

Draw and count

● Count the counters in each box. Can you see a pattern?
Draw the correct numbers of counters in each empty box to complete the pattern.

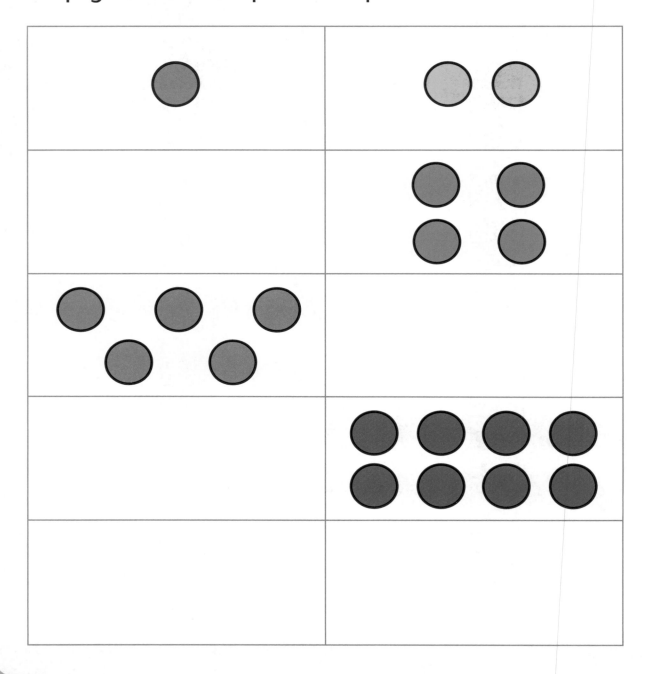

On the next page, encourage your child to find and count the objects that need to be coloured, e.g. "We have to find two houses. Can you count two? Can you count seven sails on boats?" They may not know what a buoy is — explain that boats can be tied to these at sea or in rivers.

More counting 0–10

● Look at the picture and colour:
1 dog, **2** houses, **3** people, **4** birds, **5** trees, **6** boats,
7 sails, **8** buoys, **9** butterflies and **10** fish.

Well done!
Now colour
the star.

More or fewer to 5

Tick (✓) the branch that has **more** squirrels.

Circle the squirrel that has **fewer** acorns.

On the previous page, there were boats tied up in the river. There are eight buoys but only six boats – two buoys did not have a boat. Go back and look at this page and talk about *more* and *fewer*, e.g. there were *more* buoys than boats.

More or fewer to 10

- Look at the number under each squirrel.
  Give each squirrel **2 more** acorns.

6

8

- Which squirrel has **fewer** babies?
  Draw a ring around that squirrel.

Estimating to 5

- Guess how many stars are in the sky. Write how many you think there are.

- Now count how many stars there are. Write the number.

- How many balloons do you think there are? Write the number.

- Now count how many balloons there are. Write the number.

You can practise estimating and counting with your child anytime. When you are out, you could say, 'How many swings do you think there are? Do you think it's more or less than …? Let's go and count and see if we are correct." or "How many acorns do you think are in my pile? I think there are …? Let's count to see who is correct."

Estimating to 10

- Each player needs a football.
  How many do you think you will need?

- Draw a football for each player.
  How many footballs did you draw?

- How many medals do you think there are?
  Write the number.

- Now count and colour the medals.
  Write the number.

Well done!
Now colour
the star.

Adding

● Count each group of animals.
  How many are there altogether?

   and   **=**

     and **=**

 **and**  **=**

 **and**  **=**

Count one group and then continue to count to add on the second group. This is an early introduction to addition. As your child improves, add three groups. You can use anything from pebbles to toys or pegs!

More adding

Add these numbers together and write the answer.

 +  1 add 1 = ☐

 +  3 add 2 = ☐

 +  5 add 4 = ☐

 +  5 add 5 = ☐

Well done!
Now colour
the star.

Take away

Cross out the number of sweets being taken away. How many are left?

 take away 2 =

 take away 1 =

 take away 0 =

 take away 5 =

More take away

Take away the second number.
Write the answer.

2 take away 1 = [ ]   − 1

4 take away 2 = [ ]   − 2

3 take away 3 = [ ]   − 3

5 take away 4 = [ ]   − 4

Well done!
Now colour
the star.

Counting 11

Find and circle **11** birds in the picture.

Counting 12

- **12** little ducks went swimming one day.
  Count and colour **12** ducks.

Well done!
Now colour
the star.

Counting 13

● Tick (✓) the bowl with **13** fish.

☐   ☐   ☐

● Count the fish from **1** to **13**.
Circle number **13**, the last fish.

Using a few balls of playdough, stick a piece of dry spaghetti into each ball. Put a number label by each ball. Help your child to count the correct number of small beads to match the number and thread them onto the spaghetti.

Counting 14

Join the dots to draw **14** fish in the tank.
Now count the **14** fish.

Well done!
Now colour
the star.

Counting 15

- Here are **15** butterflies and **15** flowers.
  Draw a line to join each butterfly to a flower.

Counting 16

- Count the number of minibeasts in each set.
  Tick (✓) each set that has **16** minibeasts.

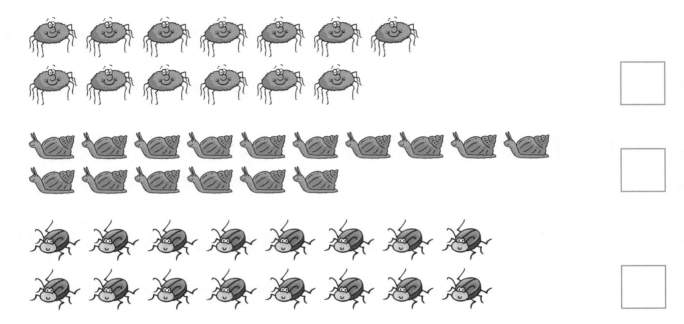

☐

☐

☐

- How many flies did the spider catch?
  Count and write the number.

☐

Counting 17

How many T-shirts are on the washing lines?
Count and write the number.
Now colour the T-shirts.

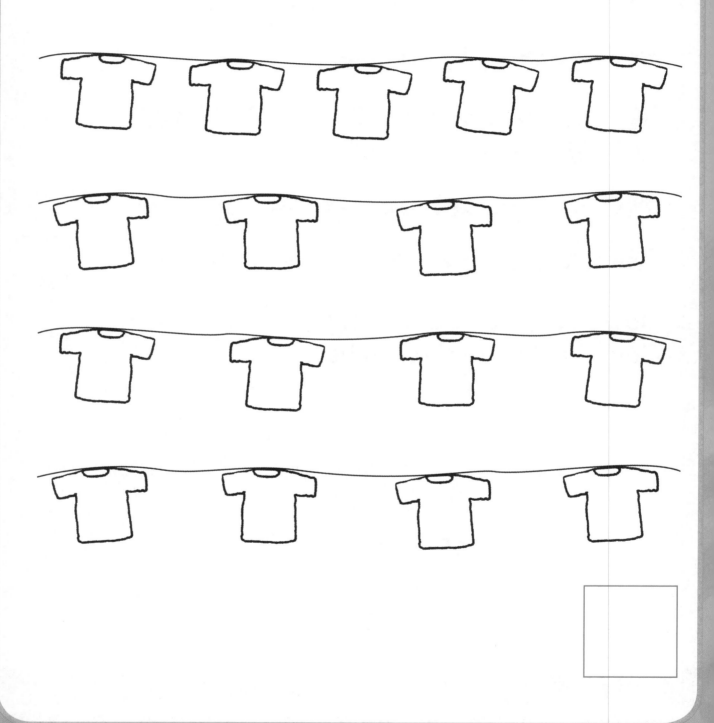

If you are hanging out your washing, incorporate some counting fun. "How many clothes are we hanging out? How many pegs will we need? Can you count some out for me?"

Counting 18

- Here are **18** hats.
  Count the hats and draw patterns on them.

- Draw a ring around the T-shirt that has **18** spots.

Counting 19

- Draw a line from the dinosaur to the set that has **19** leaves.

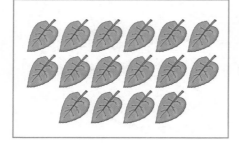 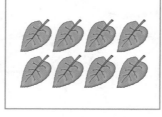 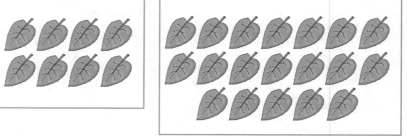

- Count each section of the dinosaur's sail. Colour the sail.

If you have lots of little dinosaurs or a set of anything else that is small, use them for counting. Roll a dice and take out that number of items. Take it in turns to do this. When all the items have been taken, count each set and see who has won the most!

Counting 20

● Count how many bony plates are on the dinosaur.
Write the answer and colour the plates.

● Trace the crocodile's teeth.
Count how many teeth it has and write the answer.

Well done!
Now colour
the star.

Counting activities

Look at the number under each group.
Draw a ring around that number of instruments.

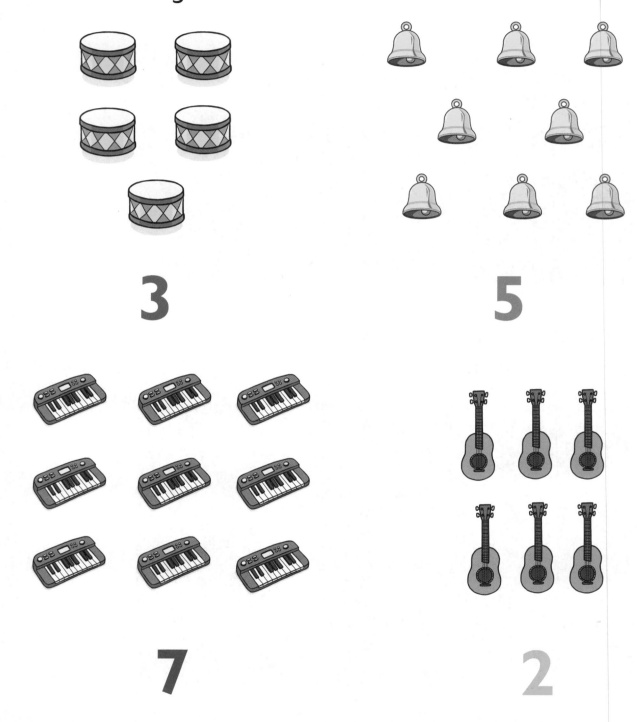

3

5

7

2

Use a group of objects to continue this exercise. Suggest a number to be removed from the group, e.g.
"Give me two spoons." or "Pick out four cars."

Counting fun

Draw a line to match each group to another group with the same number of instruments.

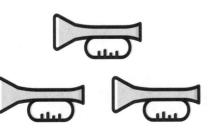

I more

Draw **I more** in each row.
Count and write how many there are now.

 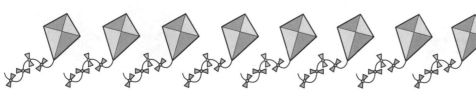   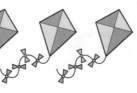 

Continue this exercise with a group of objects. Select some cars, for example, from a group or cars and say, "Can you give me one more?" Give your child a group of cars and say, "Take away one car so that you have one less."

1 less

Cross out **1** toy in each row so that there is **1 less**.
Write how many there are now.

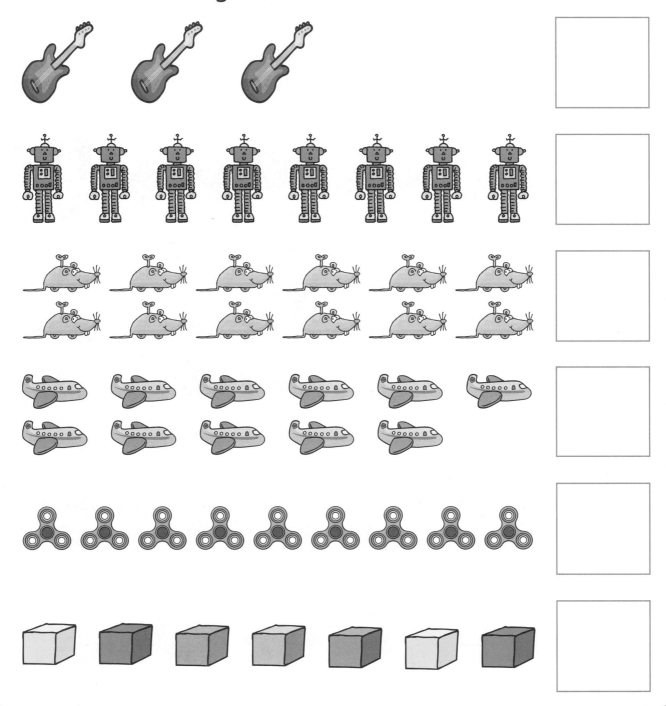

Addition

- Add the two numbers and write the answer.

6 plus 2 = ⬜  +

7 plus 5 = ⬜ +

10 plus 3 = ⬜ +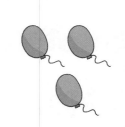

- Now add these numbers.
  You can use your fingers or small objects to help you.

3 + 3 = ⬜   6 + 1 = ⬜

4 + 1 = ⬜   1 + 1 = ⬜

Subtraction

- Take away the second number from the first.

5 minus 1 = ☐

10 minus 5 = ☐

7 minus 2 = ☐

- Draw lines to match the first number.
Cross out the number of lines that you have to take away. Write the answer.

4 − 1 = ☐
| | |/

6 − 3 = ☐

5 − 4 = ☐

7 − 2 = ☐

Making 10

- Count the items in each set.
  Write the numbers in the sum.

 + ☐ + ☐ = 10

 + ☐ + ☐ = 10

 + ☐ + ☐ = 10

 + ☐ + ☐ = 10

 + ☐ + ☐ = 10

Number bonds are not taught until Key Stage 1. However, understanding that certain combinations of two numbers will always make 10 is a life skill and will help with later learning.

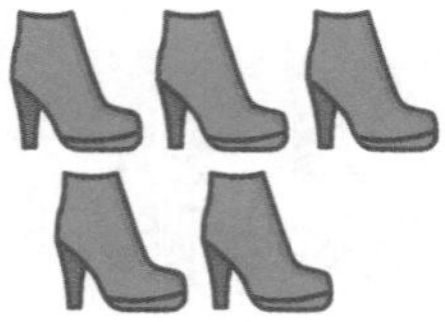

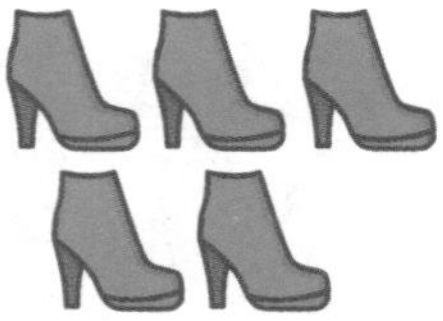

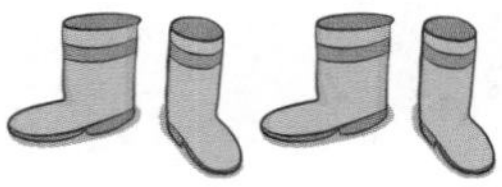

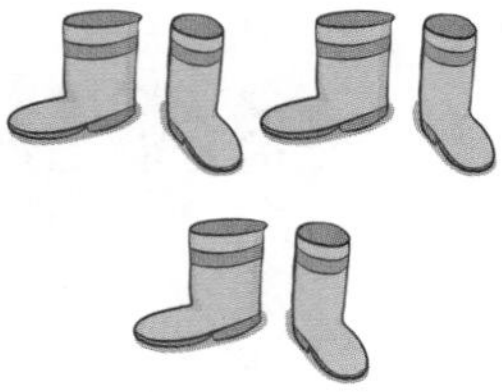

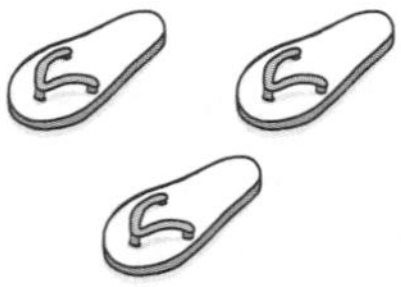

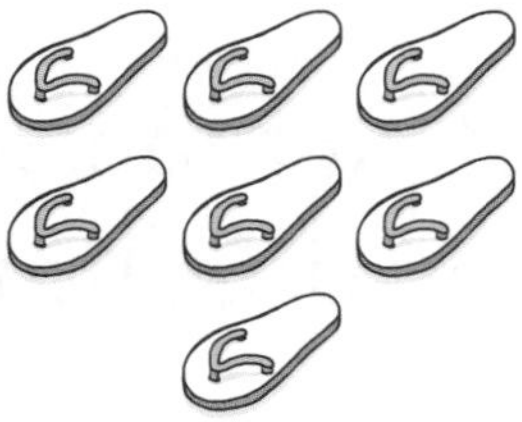

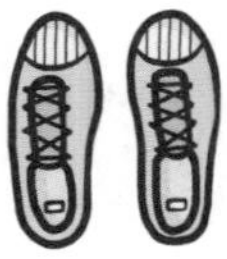

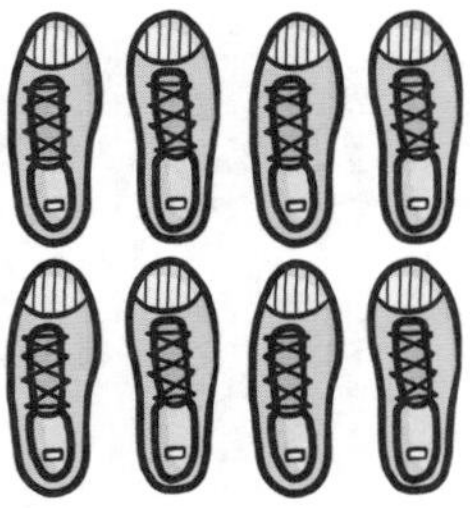

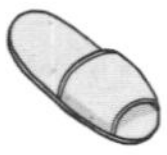

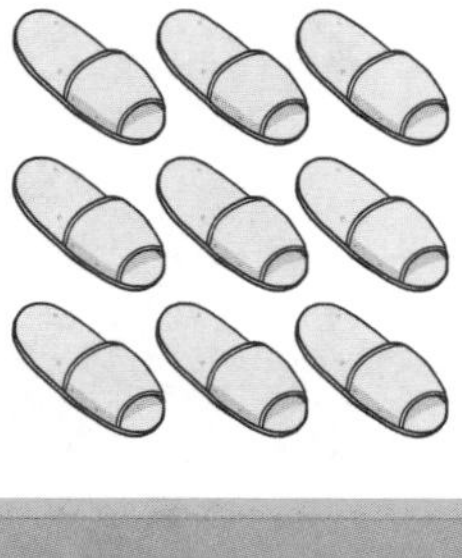

More making 10

● Draw lines to make pairs that add up to **10**.

5

9

2

7

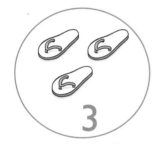

3

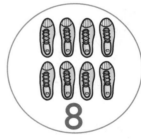

8

6

5

1

4

Well done!
Now colour
the star.

1 to 10

● How many coloured sections are there on each umbrella? Write the number.

Counting to 20 is a goal for the end of Reception, so extra help may be needed. However, it can be achieved if your child is enjoying it. On the next page, start at the red section each time and help your child to count by pointing to each section in turn.

11 to 20

- Now count the coloured sections on these umbrellas. Start at the red section each time. Write the number.

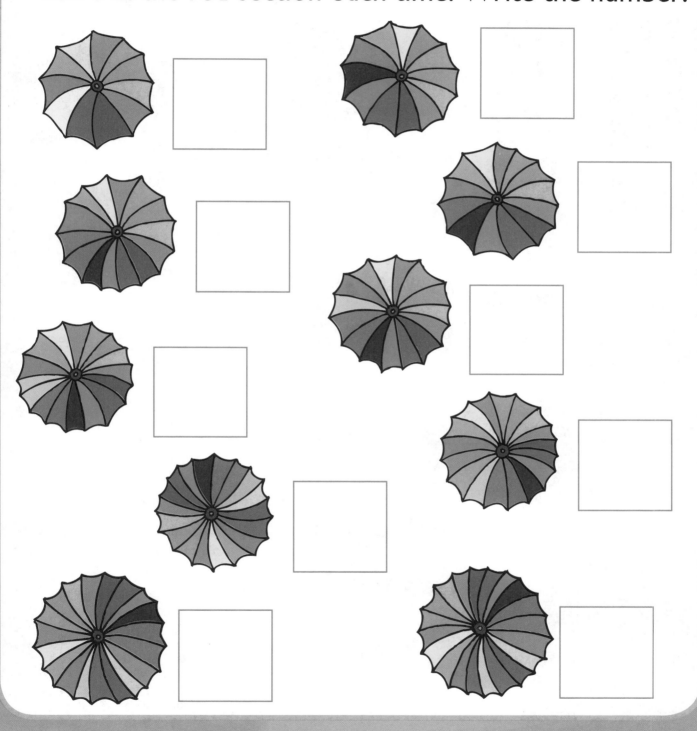

Well done!
Now colour
the star.

Answers

Page 6

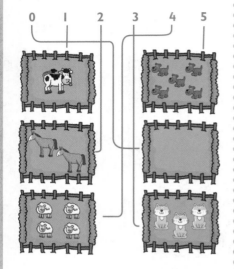

Page 7

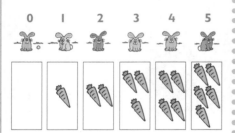

There is 1 molehill without a mole.

Page 8

The stars can be any colours.

Page 9

Page 10

The cacti can be any colour. Any 8 may be coloured.

8

Page 11

Page 12

Page 13

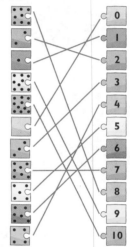

Page 14

Any colours and arrangements of counters can be used as long as the numbers are correct.

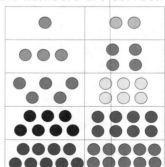

Answers

Page 15
Any colours can be used to complete the picture.

Page 16

□

 ✓

Page 17

Page 18
Child's own estimate.

4

Child's own estimate.

 5

Page 19
Child's own estimate.

 7

Child's own estimate.
Any colours may be used.

 10

Page 20
 and = 5

and = 6

and = 6

and  = 10

Page 21
+ 1 add 1 = 2

+ 3 add 2 = 5

+ 5 add 4 = 9

 +  5 add 5 = 10

Page 22
 take away 2 = 2

 take away 1 = 4

 take away 0 = 3

 take away 5 = 0

Page 23
2 take away 1 = 1   − 1

4 take away 2 = 2   − 2

3 take away 3 = 0 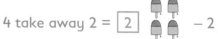 − 3

5 take away 4 = 1   − 4

Page 24

45

Answers

Page 25

Any colours can be used to complete the picture.

Page 26

Page 27

Page 28

Each butterfly can be joined to any flower.

Page 29

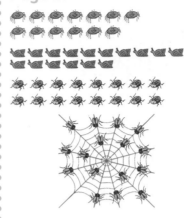

16

Page 30

Any colours can be used.

17

Page 31

Any patterns can be drawn.

Page 32

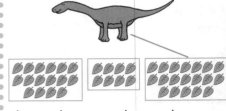

Any colours can be used.

Page 33

Any colours can be used.

20

20

Page 34

Accept one ring around the correct number of instruments in each group.

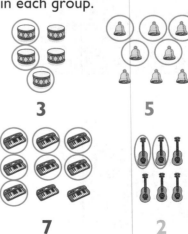

3   **5**

7   **2**

Answers

Page 35

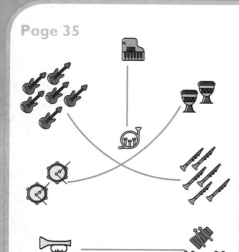

Page 36

5

6

8

10

9

Page 37

2

7

11

10

8

6

Page 38

6 plus 2 = 8

7 plus 5 = 12

10 plus 3 = 13

3 + 3 = 6   6 + 1 = 7

4 + 1 = 5   1 + 1 = 2

Page 39

5 minus 1 = 4

10 minus 5 = 5

7 minus 2 = 5

4 – 1 = 3   5 – 4 = 1

6 – 3 = 3   7 – 2 = 5

Page 40

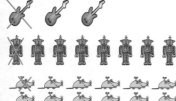 5 + 5 = 10

4 + 6 = 10

3 + 7 = 10

2 + 8 = 10

1 + 9 = 10

Page 41

5

2

3

6

1

9

7

8

5

4

Page 42

3

1

2

4

5

6

7

8

9

10

Page 43

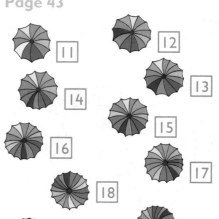

11

12

14

13

15

16

17

18

19

20

Collins Easy Learning
Certificate of Achievement

Well Done!

This certificate is awarded to ....

for successfully completing ....

Age ....   Date ....

Signed ....

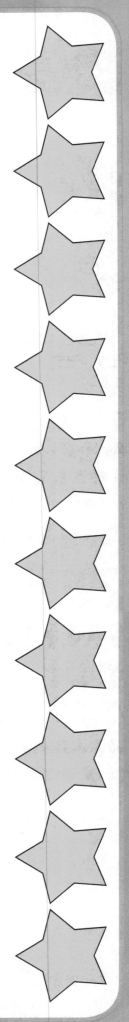